Alexander Thomas Fraser

An Historical Review of the Principal Jewish and Christian

Sites at Jerusalem

Alexander Thomas Fraser

An Historical Review of the Principal Jewish and Christian Sites at Jerusalem

ISBN/EAN: 9783337038168

Printed in Europe, USA, Canada, Australia, Japan

Cover: Foto ©Lupo / pixelio.de

More available books at **www.hansebooks.com**

AN HISTORICAL REVIEW

OF THE PRINCIPAL

JEWISH AND CHRISTIAN SITES

AT JERUSALEM.

BY

MAJOR FRASER, R.E.

LONDON:

PUBLISHED BY EDWARD STANFORD, 55, CHARING CROSS.

1881.

Price One Shilling.

AN HISTORICAL REVIEW

OF THE PRINCIPAL

JEWISH AND CHRISTIAN SITES AT JERUSALEM.

BY

MAJOR FRASER, R.E.

LONDON:

PUBLISHED BY EDWARD STANFORD, 55, CHARING CROSS.

1881.

AN HISTORICAL REVIEW

PRINCIPAL JEWISH AND CHRISTIAN SITES AT JERUSALEM.

IF an equal surface-projection of the earth's surface be taken, Jerusalem will be found nearly, if not quite, at the centre of the whole of the dry land on the globe. It also, by the assenting homage of the great majority of the human race, holds the same position in the universe of mundane thought and destiny. Why this is the case, and in what respects the site of this city differs from every other place on the face of the earth, are unknown. Perhaps the only other problem of the kind, though of far less importance in comparison, is the exact situation of the Garden of Eden. The earliest possessors of the site of Jerusalem were the Jebusites, sprung from Canaan, the youngest son of Ham, who appear to have been unaware of anything except the strength of its position. Long before Jerusalem became the city of the Jebusites, the faint breathing of tradition speaks of a temple in the rock, with an entrance by nine porches, and supported by two pillars with a perpendicular depth of eighty-one feet, in which was deposited an equilateral triangle of gold enriched with precious

1—2

stones encrusted on an agate, and containing an all-powerful name; the whole reposing on a pedestal of white marble, with which the name of Enoch is connected. Then succeeds a passing mention of Melchizedek, king of Salem and also priest, and who must have lived about 1918 B.C. at Jerusalem. Shortly after, Abram, residing in Philistia, went towards the land of Moriah, and saw the place at which he was to sacrifice his son, on the third day, far off. We are told he called the place " Jehovah Jireh," and that it was in the Mount of God. It is commonly accepted to have been upon the heights enclosed in the Haram Area. Tradition goes that Shem, son of Noah, settled at Jerusalem, and imparted specially important knowledge to Abram, which, as he died about 1824 B.C., might easily have been the case. The probability is that a priesthood was founded by Melchizedek, which the patriarch then continued. But there is a mere glimpse afforded at this period of an age which has otherwise left its mark in the profound depths of the immediate neighbourhood.

Then there succeeds another patriarchal system, the Exodus, and the wanderings, and we have Jerusalem assigned 1444 B.C. to the tribe of Benjamin. Even in the time of David the Jebusites had still their stronghold, not, we may be sure, on the dry sides of the Haram position, but opposite, on the declivity looking north-cast, which carries the modern town.

The slumbering importance of Mount Moriah was first disclosed upon the pestilence succeeding the numbering of the people. The angel stood, the sacrifice was offered, and David made a purchase of the threshing-floor of Araunah. To know that it was such a floor is no great help in determining the site. For threshing by oxen in the East, all that is required is a flat circle of rock or beaten earth, about a dozen feet in diameter, the more sheltered the better if there is a saddle or eminence close by, on which winnowing of the grain can take place. Solomon was divinely authorised to build the Temple, and the construction was managed with the assistance of Canaanitish hands. The Syrian workmen were employed in the masonry, and Hiram, a widow's son of the tribe of Naphtali, and whose father was of Tyre, superintended the brazen castings. In most books taken up by chance, whether of history or sacred topography, it will be found stated that Solomon's Temple was built on the top of Mount Moriah, in the place where now stands the Mosque of Omar. The locality is so plausible at first sight, that it has been readily accepted, until now it is most difficult to procure attention to the want of evidence or proof upon which the statement rests. The work was of great magnitude. Solomon raised a levy of 30,000 men from Israel, and sent them 10,000 a month by courses to Lebanon, to fell and transport timber, keeping a course two months at home. Adoniram was over the levy. He had, including the Syrians, 70,000 people carrying loads, and 80,000 masons under 3300 superintendents. Immense hewn stones were prepared at the quarries, so that no sound of iron tools could be heard during building. Josephus gives an account of Solomon's Temple (Ant. b. viii. chap. iii. sec. 2), which has caused difficulty. But it seems intelligible enough if the plain narration is followed. He says, " The king laid the foundations of the Temple very deep in the ground." To begin with, this cannot refer to the site of the so-called Mosque of Omar, where there is hard rock near the surface, and least occasion to dig deep. The Temple must therefore have stood on some place where there was a thickness of earth, and nearer the base than the top of Moriah. The dimensions Josephus gives are: height 60 cubits, length 60 cubits, and breadth 20 cubits, to roof. There was another structure over it of equal dimensions, so that the whole height was 120 cubits. There was a porch 20 cubits long by 12 cubits wide, and the same height as the Temple. The dimensions in feet would of course depend upon whether the cubit of a man, some 18 inches, or the sacred cubit of 25·05 inches were adopted. In the one case the height would be $218\frac{2}{3}$ feet, otherwise $250\frac{1}{2}$ feet. However, Solomon's Temple was at its first appearance a tower of from 218 to 250 feet in total height, built, as I believe, of two storeys, in the south-west corner of the present Haram Area, and over that part of it which has not been yet explored, but near which Julian the Apostate's workmen came upon the globes of flame. Just in front must have stood the threshing-floor of Araunah, on which the altar was built. This explains how the porch had such an altitude of 120 cubits. Little is known about the arrangements of Solomon's Temple. There were upper chambers inlaid with gold, no doubt for quasi-secular uses, protected by the sanctity of the rooms of worship below. It is most likely that the rest of the hill was irregularly laid out; some of it stepped off in terraces, other parts built upon and filled up, while Josephus states (Ann., b. xv. chap. ii.), "The hill was a rocky ascent that declined by degrees to the east side of the city, till it came to an elevated level. This hill Solomon had compassed with a wall." The buildings must have looked singularly picturesque, and were the expression of a new order of government upon earth, but quite different from the Temple and its courts and palace at a later period. Solomon had, according to Josephus, encompassed the hill with a wall by divine revelation, and this no doubt equally guided the position of the Temple foundations. A great expense was incurred in laying these, which might have been easily saved by moving some yards away; but the place was chosen with an object, with which we are not directly made aware.

Solomon's Temple lasted till the captivity took place, 588 B.C., under Nebuchadnezzar, when the Chaldeans broke up the fine castings made by Hiram, and took the vessels of the Temple to Babylon. They burnt the house of Jehovah, brake down the walls of Jerusalem, and destroyed the palaces and goodly contents, carrying off the treasures. It may be easily understood that a temple in the form of a lofty tower would, when once it fell into the possession of the Babylonians, lined as it was with cedar, burn like a furnace, and be soon reduced to complete ruin.

So the city and Temple remained till 536 B.C., when the spirit of Cyrus, king of Persia, was stirred to proclaim the rebuilding of the house of the Lord; and we are told the people gathered themselves together as one man to Jerusalem. The same process had to be gone through as at the original instauration. Jeshua and Zerubbabel were over the work, Syrians again brought timber from Lebanon, and masons and carpenters were employed by the Levites. There was no question regarding the site on this occasion; because the priests and ancients had seen Solomon's Temple standing. The decree of Cyrus contained the specification for the Temple, but the Temple was actually built under Darius, who found the decree of Cyrus in the house of the Rolls. The expense was originally to have been defrayed by the king's treasury. There is no certainty that the limits prescribed by Cyrus were not exceeded, for the people helped the

work, and the restoration of the walls of the city was
divided amongst the leaders. Probably the Temple of
Zerubbabel was a close reproduction of that of Solomon,
then fresh in recollection. Hasty work or defective orna-
mentation was remedied about 301 B.C., when Simon, the
high-priest in the reign of Ptolemy Lagus, added to the
walls of Jerusalem.

In 168 B.C. Antiochus Epiphanes profaned and despoiled
the Temple, but did not destroy it, and its worship
and influence were restored by Judas Maccabæus. The
Jewish polity continued unbroken within the Temple
enclosure under the Asmonean princes, until, the throne
being vacant, the Pharisees proposed one candidate and the
army another.

A conflict followed, and the Romans interfered. Pompey
decided in favour of Hyrcanus II., the candidate of the
Pharisees, and in 63 B.C. captured Jerusalem. Josephus
narrates ("Wars," book i. chap. vii.) : "Pompey saw the
walls were firm, the valley before the walls terrible; the
Temple which was within that valley was encompassed
with a very strong wall, so that if the city were taken the
Temple would be a second place of refuge for the enemy to
retire into." If this passage is read backwards it conveys a
clearer meaning, because this order goes from the abstract to
the concrete. The Temple was a second refuge if the city
were taken; it had a strong wall; the Temple was within a
valley (the Tyropœon and not on the top of Mount Moriah),
and the valley to the south of the Temple and city walls
terrible then, as is the case now, but more conspicuously so
because not filled with rubbish. Josephus proceeds : "Aris-
tobulus's party retired into the Temple, and cut off the
communication between the Temple and city, by breaking

down the bridge that joined them together. But Pompey himself filled up the ditch that was on the north side of the Temple, and the entire valley also." For an explanation of this passage we have only to turn to Wilson and Warren's " Recovery of Jerusalem." The remains of this bridge exist to the present hour. No one can also look at the longitudinal section of the Haram Area from south to north without recognising the hollow which is now partly closed with earth, and in part occupied by the pool of Bethesda, as the ditch on the north side to which Josephus refers. It was, of course, not of the full section all through, but connected itself with the head of the Tyropœon valley by a depressed saddle, if not by a steep artificial cut. When the Temple was taken, Pompey entered the Holy of Holies, and it was remarked that from this time he ceased to be successful. But he rifled none of the treasure, and cleansing the Temple, directed the resumption of its services. Crassus, the prefect of Syria, was in 51 B.C. not so moderate. He plundered the wealth that Pompey left untouched, stated to have amounted to two millions sterling. The defeat of Crassus by the Parthians hastened the strife between Cæsar and Pompey ; the Rubicon was crossed, and the defeat at Pharsalia ended the consulate.

On the ides (15th) of March, 44 B.C., Julius Cæsar came by his tragic death. The year before he had attained, in all except name, absolute sovereignty over the Roman Empire, and had just convened a meeting of the Senate to obtain the imperial title. The conspiracy against him was an outburst of suppressed republican feeling, and for some time the government was conducted by the joint presidency of Augustus and Antony. The battle of Philippi, 42 B.C., ended republicanism in Rome, and confirmed their authority.

With his colleague's consent, Antony, 40 B.C., made Herod king of Judea. Antipas, the grandfather of Herod, was an Idumean, and governor of Idumea under Jannæus; who, being a Sadducee, was opposed by the rival Jewish faction· The support of the Pharisees contributed to the elevation of an Idumean over Jerusalem, and the sceptre departed from Judah. It has never been clearly related how the gap occurred that made way for Herod's Temple. That of Zerubbabel emerges with apparently a fabric in material respects undamaged, after attack and pillage on these several occasions, and there is no distinct record of Herod having found a mere heap of ruined buildings to invite restoration. To trace the condition of Zerubbabel's Temple, it is necessary to go back to 168 B.C., when Antiochus Epiphanes caused the discontinuance of the daily sacrifice. At that period (1 Mac. i.) " her sanctuary was laid waste like a wilderness ;" but this may not imply more than removal of the Jewish ornaments and a dismantling of the interior. But the orders of Antiochus were " that they should follow laws strange to the land ;" the religion of the Jews was suppressed, " overseers were appointed over the people," the Temple of Jerusalem was dedicated to Jupiter, an image set up, and sacrifices offered on a new altar. It was the practice to erect the altars for such strange rites upon the tops rather than sides of hills, and the description given in 1 Macc. iv. of the condition of the Temple when Judas Maccabæus proceeded to cleanse it, " all the host assembled themselves together and went up into Mount Sion. And they saw the sanctuary desolate, and the altar profaned, and the gates burnt down, and shrubs growing in the courts as in a forest or in one of the mountains, yea, and the priests' chambers pulled down," is inconsistent with the decorous cult even of Olympian Jupiter. The altar of the Jews rendered unclean, the Temple was deserted; and if a shrine were not built, the summit of Moriah can nevertheless be conceived smoking with unblessed fires.

Towards the close of 165 B.C. the stones were clinked together, the Jews rekindled their daily sacrifice on a fresh altar, and Judas Maccabæus fortified the Temple, after having built up the sanctuary and the inner parts of the house. The celebration of the event lasted eight days, and was annually repeated as the feast of the dedication, even in the Temple of Herod. The Maccabæan rulers added to the fortifications of Jerusalem, and national freedom was restored under Simon, in 143 B.C., high-priest, governor and leader of the Jews. A rather remarkable passage with reference to him occurs in 1 Mac. xiv.: "For in his days things prospered in his hands, so that the heathen were taken out of their country, and they also which were in the city of David in Jerusalem, who had made themselves a tower, out of which they issued and polluted all about the sanctuary, and did much hurt in the holy places; but he placed Jews therein, and fortified it for the safety of the country and the city, and raised up the walls of Jerusalem." As Simon is narrated to have raised a stronghold near the Temple for his own residence, which after became the tower of Antonia, it is not at all improbable that the position was at any rate thus marked by the tower of the heathen in the city of David, which must have closely adjoined.

In 109 B.C. Hyrcanus, Simon's son, built the Castle Baris, a fortress on the north-west corner of the Temple courts, square in shape, and which was afterwards expanded into the Tower Antonio. The history of Zerubbabel's Temple is, up to this point, epitomised by Milton:

" Returned from Babylon by leave of, kings,
  Their lords whom God disposed, the house of God
  They first re-edify ; and for awhile
  In mean estate live moderate ; till grown
  In wealth and multitude, factious they grow ;
  But first among the priests dissension springs,
  Men who attend the altar, and should most
  Endeavour peace : their strife pollution brings
  Upon the temple itself ; at last they seize
  The sceptre, and regard not David's sons ;
  Then lose it to a stranger."

Nothing is stated with regard to the condition of this
Temple between 109 B.C. and 17 B.C., when Herod com-
menced the more splendid edifice, which has sunk in a more
profound oblivion. But it may be readily conjectured that
a building completed 515 B.C., gutted in 168 B.C., hastily re-
habilitated three years afterwards, and of whose repairs
nothing is heard after 109 B.C., had grown dilapidated and
unsuitable by 37 B.C., when Herod entered upon his government. Being an Idumean, he could not hold the office of
high-priest ; but he could give a permanent tone to the
mode of conducting the services of the Jews by causing the
erection of a new Temple. It is evident the old structure was still existing, because the Jews expressed a fear
that if he pulled it down they would have none in its place;
but Herod gave a pledge that he would not begin demolition
till materials had been collected. Josephus describes
Herod's procedure in building the new Temple in a way
that is easily followed (Ant. b. xv., chap. ii.) : "Herod took
away the old foundations and laid others, and erected the
Temple there, 100 cubits long and 20 cubits high, which fell
down on the sinking of their foundation." It is incredible
that this construction was on the hard crown of Moriah
where the Dome of the Rock now stands. It must have

been on much lower and far more slippery ground. Josephus proceeds : "Temple stones were white and strong, 25 cubits long, 8 cubits high, and 12 cubits broad." "The whole structure, as also the structure of the Royal Cloister, was on each side much lower, but the middle was much higher, till visible to those living in the country for a great many furlongs, but chiefly to those who lived opposite." It will be found on referring to the photographs of the Ordnance Survey of Jerusalem, that if Herod's Temple be assigned to a position south of the Sakhra, it must have been very conspicuous from the north and west sides of Jerusalem, and in the most advantageous grouping with the spread of the city, when viewed across the valley from the east, or over the Tyropœon from the roofs and windows of the town. Josephus continues : "The neck was rocky, anticlinal, quietly slanted with (πρoς) the eastern members of the city to a summit point. This ridge Solomon had compassed with a wall," or in Greek, with interlinear translation :

"*Τουτον ὁ πρωτος ημων βασιλευς Σαλομων*
    This, the foremost of us King   Solomon,
και   επιφροσυνην του Θεου μεγαλαις εργασιαις
and the wise plan of God, with great operations,
απετειχιζεν ανωθεν   τα περι την ακραν
walled off, to be above those round the summit."

The passage has given rise to uncertainty whether in what follows Josephus is not recounting the dimensions of Solomon's Temple; but it is evidently a parenthesis, and Josephus resumes the doings of Herod. When the wording however is closely examined, it is very applicable to the tower-temple built by Solomon, from a low site in the south-west corner of the Haram Area, so high as to over-look the crown of the ridge.

As to Herod, "He also built a wall till the largeness of the square edifice was immense, and its altitude: επιτειχιζε δε κατωθεν υπο της ριξης αρχομενος he walled over also from below under the spur that com-manded, etc." The term, an immense square edifice, cannot relate to Solomon's Temple, whose dimensions and mode of construction were, as described, very different. Josephus continues : "This work was joined together as part of the hill itself to the very top ; he made it level on the external surface, and a smooth level also. This hill was walled all round and in compass four stadia, the distance of each angle containing a stadium (or 604·35 feet)." Then in Greek :

"Εσωτερω δε τουτου και παρ αυτην την ακραν,
Within also   it,   and towards the very summit,

αλλο τειχος ανω λιθινον περιθει, κατα μεν
another wall tier of stone runs round, on   in fact

εωαν ραχιν των τειχων, και την προ τουτων
the east slope   of the walls,   and fronting these

φαραγγα φοβεραν το τε ιερον εντος της φαραγγος.
a hiatus formidable, and the Temple within the hiatus.

Αυτο δε κατα το προσαρκτιον κλιμα την του
That also on the northern   incline on that of the

ταφρου εχου και την φαραγγα πασαν υλην
sepulchre he heaped; and the entire hiatus material

συμφορουσαν της δυναμεως.
brought   by the labour.

The meaning of this somewhat involved passage seems to be that Herod, dealing with the same features of the ground as Solomon, began to build an immense square revetement, just to south-west of the Sakhra, then probably clad with soil, and barely showing above the surface. It was carried up some height, perhaps eighty feet, at one corner, where there was a string-course, or some mural line; and then came a second tier of walls, the interior, about 600 feet square, being at first, while the work was in progress, a vast hiatus. In the middle of this the Temple foundations were relaid, and the superstructure brought up to the level of the revetement outside, in the seclusion that was desirable for building not only the sanctuary, but the treasure-house of the nation. There appears to have been a place of sepulture on the north, on which, and within the walled limits, he naturally deposited the materials while the work was in progress. The revetement was made not to join on to the top of Moriah, but so solid as to resemble the rock itself right up to the top of the walls; and on the north was built a citadel square and strong.

In this way Herod's Temple foundation stood out a square and protected block of masonry, in the most picturesque spot that could have been chosen on Zion; and the Temple itself stood with its cloisters and courts on this basement, indicative of a centre for the Jewish polity that was originally marked, not by the summit of the ridge, but by divine command. The sections which are given of the Haram Area in the "Recovery of Jerusalem," impart the idea that the basement of Herod's Temple was of a greater height than generally supposed; for the shingle which now fills up the old bed of the Kedron, and half obliterates the Tyropœon Valley on the opposite side, has evidently been poured or drawn down from a higher level, or from being, in fact, much of it at one time the filling of the basement of the terrace. Before the invention of gunpowder, a steep

cyclopean square of masonry, 150 feet in average elevation, would have presented a nearly insuperable obstacle
to attack, from an enemy operating in a mountainous
country, possessing few supplies. The north was the
feeblest side; but the summit of Moriah, evidently from
Ezekiel (xliii. 7, 8) a remarkable passage, contained the
burial-place of the kings, and was in close connection with
the Temple, perhaps joined by a terrace; and the whole of
the open space to the north was elaborately defended by sets
of walls. It would not be difficult, in a model of the city of
David and hill of Zion, to reproduce the entire system of
the Temple and outworks from the elaborate information
accumulated by the Ordnance Survey, and the researches of
Mr. James Fergusson and others, with some approach to
fidelity. But this work, although admirably attempted,
cannot be said to have been yet conclusively accomplished.
Upon the supposition that the area of the Temple mea-
sured a stadium, or 604·35 feet, and that the cubit of
Josephus and the Talmud was 1.824 feet, the cubit of a man
of Deut. iii. 11, the ground-plan minutely worked out by
Mr. Fergusson leaves little to be desired in the present state
of the question. It brings the holy of holies, and the altar,
opposite where the Jews to this day congregate to bewail
the misfortunes of their tribes, and the alienation of their
holy house. Antonia is placed at the north-east corner,
projecting half out, and half inside, the Temple wall, and
continued in its outbuildings, over the ruins discovered by
Captain Warren. But the position of the tower is open to
the objection that its demolition would not exactly fit the
weird tradition that the Temple was not to be taken till it
became four square. The Antonia was an old tower already
built when the foundations of Herod's Temple were being

laid ; and it is unlikely that the massive revetements would
be cut short, and abut against an old wall, but more reason-
able to suppose that the Antonia was itself built against in
such a manner, that the Temple really stood out square
when the tower was demolished. The point is, however, of
minor importance. A very symmetrical restoration of the
ground-plan of the Temple of Herod is contained in the
finely-executed work of the Marquis de Vogué. The inducement to place the altar of the Jews on the top of the Sakhra
arises from a difficulty experienced in rejecting the Golden
Gateway as being connected with the Temple, its principal
entrance, and beautiful gate. The authorisation is a somewhat vague passage in Josephus, " Wars," i. chap. 21, 1 :

Τον ναον επεσνευασε και την περι αυτον ανετειχισατο
The temple   he heaped, and   round this   walled

χωραν της   ουσης   διπλασιαν.
a space of what was, the double.

There is no way of determining the exact limits of the
walled enclosure which Herod doubled.

From such account as we possess of Zerubbabel's Temple,
the actual court must have been of irregular outline, because on sloping ground ; and as the area of a figure nearly
four-sided is as the square of the side, 420 feet increased to
600 feet would give the requisite double space, and at the
same time afford the widest extent that could be reasonably
claimed for the Temple of Zerubbabel. The plan, as restored
by De Vogué, has undeniable nobility of conception, and
thorough command of the key of the position. But if eleva-
tions were drawn, and still more in perspective view, the
Haram constructions of Herod's time form a feeble group
spread over such a wide area as the walled enclosure which

De Vogué assumes. There were also several walls defend-
ing the ground immediately adjacent to the northern side
of the Temple, and it is therefore not quite certain to
what exterior limits Josephus referred. Question on the
matter seems to arise, more than anything, from the complete absence of any visible mark of the spot on which
the altar stood. There is an unresting feeling of obligation
to find in the Haram Area some indication of the kind, and
the prominent feature of the Sakhra is seized for want of
better. But there is nothing in the conditions of the altar
to connect its position with any extruding peak. It had its
origin in the sacrificial heap raised on a threshing-floor. For
the purposes of the Temple an elevated base had to be built
33 cubits, or about 60 feet square; and if Josephus's numbers
be taken rising 24 feet above the level of the threshing-floor,
with the real altar of unhewn stones some 3 feet more,
and almost 22 feet square. There was a drain, according to
the Talmud, which carried the blood of the sacrifices away
from the top of the base to the Kedron. But there was no
indisputable natural feature to denote the site, except such
as attaches necessarily to the threshing-floor of Araunah.
The altar site is a point in the Temple area to be indirectly
fixed, rather than by referring, except collaterally, to the
face of the rock. It is by no means certain that in the
vicissitudes the Temple and Mount Zion underwent before
the time of Herod, sacrifices of a different kind to those of
the Levitical law had not been offered upon a high place,
and a desecrated altar. The façade of Herod's Temple pre-
sents more difficulty than the ground-plan. Josephus seems
grossly to exaggerate the heights; and yet it is undesirable
to reject the measurements of an eye-witness of the edifice
for those of the rabbis of the Talmud. In preparing eleva-

tions to test the imperfect description of Josephus, it is
necessary to prepare some sort of probable cross-section,
and to do this sections are required to a natural scale of the
surface of the Haram Area at the level at which the foundations were laid by Herod. One is wanted longitudinally
through the Huldah Gate, north and south ; and the other
at right angles to this, or from the modern town to the
Kedron, nearly across the wailing-place of the Jews. The
last-named is of course not at present obtainable, and the
exact conformation of the rock is a matter of absolute con-
jecture. But if the information were procurable, the revetements would have first to be laid down their 600 feet square,
as it was already described Herod built them, recollecting
that from the rubbish accumulated, the platform containing
the Stoa Basilica and the principal courts was in all likeli-
hood quite as high as, and for some part of the north en-
virons detached from, the Sakhra. It was made soil within
the Temple enclosure, and natural ground for the most part
outside. While attempting to restore the design of the
Temple itself, the heights would have to be reckoned from
the low level of the hillside, and not from the platform
formed by filling up the interior of the square. This mode
of proceeding gives ample scope as far as height is con-
cerned. A crypt, with store and treasure rooms, is at once
suggested. Then the storeys of chambers can be arranged
in tiers so as to make a priest's cell 11 or 12 feet high, or at
most 17½ feet, unless there is some other mode of arrange-
ment. For the offices of the Temple, many may well have
been in a sunk court running round three sides of the
central building, with the several courts rising in a grada-
tion that must be determined with reference to the level of .
the altar, which stood a few feet comparatively above the

2—2

environing terrace upon the south side of which the Corinthian columns of the Stoa Basilica displayed their arresting splendour. In form the façade of the Temple may have borne a resemblance to Mr. Fergusson's elaborate reproduction in his "Temples of the Jews." The roof had perhaps even a higher pitch, and either partially or wholly gilt would have been ablaze in the morning and evening sun. Reaching, as it appears to have done, from ground to roof, the porch must have formed a special feature, disclosing through its hangings some of the magnificence of the interior. The whole edifice may have had also a specific Jewish character, denoting the Temple to be a national monument.

The turning-point in Jewish history was now reached, and the Temple and its sacrificial rites were superseded by the Christian dispensation. It is memorable that the accusation preferred against our Lord was on the subject of the Temple, and making nothing of this, the high-priest directed the final personal inquiry, Σὺ εἶ ὁ Χρίστος, ὁ υἱος τοῦ εὐλογητοῦ, and obtained the plain answer, and a prediction that is yet unfulfilled. At that time doubtless Moriah was better covered with soil, and all to the north of the Temple wall was in part laid out in gardens, and much of it an open green. The whole Haram Area was probably enclosed with a wall sufficiently high to protect the Temple, but by no means so lofty as the square revetement. There was access from the town to the country past the Antonia and across the Haram Area to a gateway that may be supposed to have stood on the foundations of the existing Golden Gate. The scene of the Crucifixion must, then, have been without the gate and therefore it is likely at no great distance from its eastern face. The ground is now so much altered that it is impossible to infer from the mere surface appearance what was the situation of Golgotha. If it can be shown that the old wall at Jerusalem bent at right angles, following the street which sharply marks such a course on the modern map, it is just as easy to place Golgotha where it is shown in the Church of the Holy Sepulchre, but that the tomb prepared by Joseph of Arimathea for himself should have been just opposite and within 190 feet of the place of public execution is difficult to credit. There is perhaps no regret so poignant as that caused by the discovery that extremities have been proceeded to under a mistake. The leaders of the Jews found themselves only opposed by the feeble resistance of matter to the infinite when they had expected supernatural strength; the tie was severed, and their national welfare for ages shattered. Herod's Temple was a silent witness of the rash attempt which left its lesson to be the strength of Christendom, and the most formidable scientific problem that concerns the race of man unsolved. The Temple continued to rally round it the expiring forces of Judaism, and the holy places to the present time involve more in their mere situation than at first sight appears. Revolting against the oppression of the Romans, with hopes from the befriending aid that audibly left the Temple of Herod for ever, the Jews were besieged in 70 A.D. by Titus, and the building was consumed by fire.

The popular impression that the rituals of the Romans were in honour of creations of a childish imagination, can only be held by those unacquainted with the bearings of the subject. There can be little question that some at least of the sites on Mount Moriah, either the altar of Abraham or that of David, and the sites of the three temples, those of Herod, Zerubbabel, and Solomon, had an interest for the Pontifex Maximus and the Romans. Their power held the

Haram Area, as the Turks do at the present moment, against dreaded opposition. The Jewish tradition (Allen's "Modern Judaism") is that the descendants of Esau spread from Seir, and, increasing in numbers after the Assyrians and Babylonians had overthrown the republic of the twelve tribes, passed over and subjugated Italy, founded Rome, and, recrossing under Titus, destroyed the Temple. From the number of Jews in the neighbourhood of the Holy City in the time of Hadrian, 131 A.D., it is certain that the Romans were satisfied, so long as they garrisoned the Haram, and prevented them rebuilding their Temple. They no doubt lived and traded in the opposite city, while the Christians were growing into a compact body in unobserved retreats amongst them, under the protection of Roman impartiality. Hadrian had been initiated into the Eleusinian mysteries at Athens, and gave orders for the rebuilding of Jerusalem as Ælia Capitolina. The consequence was an insurrection of the Jews, which was put down ; and their race was entirely banished from Jerusalem. " Ubi quondam erat templum et religio Dei, ibi Hadriani statua et Jovis idolum collocatum est." (Valerius, vol. iv. p. 37.)

Such was the state of the Haram Area in 136 A.D. The attention of Hadrian was drawn to the Jews, while he tolerated the Christians. It was an obvious measure of safe policy to supplant the Temple worship by the Roman rites ; and symbolise in the two statues the combined pontifical and imperial power. But it is almost certain that the Sakhra, held afterwards in so rooted a veneration, would have at that time commanded some regard. In the absence of a specific account it is to be presumed that the oriental inhabitants, nominally Christian, maintained its historic sanctity by at least some formal enclosure ; and that they had one or more places in the town for their religious celebrations and worship. But there were many Jews among the Christians ; and a suspicion of Judaism began to attack, if it did not actually invade, the Christian body. The Church of Jerusalem split up, in 150 A.D., into Christians and Nazarenes. The Nazarenes were mere Jews, except that they believed in Christ. We lose sight all this time of the sect of the Essenes, who no doubt, with others of kindred opinions, kept alive many of the traditions connected inseparably with the contents and situation of the Haram Area.

The pressure of Christianity was felt by the emperors in their pontifical seats, and gave rise to persecutions, the records of which are fragmentary and dissatisfying. Under Severus, Maximin, and Valerian, the Christians experienced interdict and massacre. But the invasion of his family by the proscribed doctrine evoked a curious edict of Diocletian, 282 A.D., for the suppression of the Christians and the demolition of their churches ; a process which Galerius continued in the East. The probability is that it was at this time the temple of Venus was erected over the site of the Holy Sepulchre, and the tomb itself obliterated from the rock. "It appears from the earliest age of Christianity," observes De Vogué, "Les Églises de la Sainte Terre," "that there existed a number of edifices consecrated to the Holy Mysteries. It is evident that these modest chapels were concealed with care, and were plain oratories over the holy places. History mentions them not, only talks of the temples of Venus and Adonis, which for two centuries fixed the place of Golgotha and Bethlehem."

The mention of Diocletian brings us down to Constantine.

They were contemporary, and also correspondents. Learning the nearness of the invisible, Constantine was at complete variance with Diocletian in his opinion of Christianity. He wrote in the following strain to Diocletian:

Ἀλλὰ τὸ μὲν πάθημα ἐκείνου ὑπὸ τῶν προφητῶν ἤδη προκεκήρυκτο δὲ καὶ ἡ σωματικὴ γεννήσις αὐτοῦ.

"And Babylon will be wasted, says the prophet, and Memphis lies in ruins; not by missiles but by prayer."

The burning question of the Temple, but not that made with hands, naturally reappeared, in a simple and uninformed age, to agitate the Church. Quite beside the articles of faith, in its kind a profound, but from any other point of view superfluous, question of science, sects arose, from the impossibility of explaining the physical facts attending the incarnation, death, resurrection, and ascension of Jesus Christ. To be in a position to do this, the mind would have to form an idea and explain the conditions of physical existence. But early Christendom attempted a task for which they had not acquired the scientific data, and, stumbling here and there upon an ill-expressed truth— for the analytical grasp of the human intellect often goes beyond ascertained facts—divided irreconcilably upon the scientific question. According to Neander, there were at first two theological systems in the Church. One was theistic in tendency, which distinguished the Son of God from all created beings, and maintained what is termed the " unity of essence." This became the system of the Western Empire, and practically supported the principle of monarchy in civil government. In the East the " unity of essence" was combated, and a theory of " emanation" propounded; so as to make a greater distinction between the personality of God and the Son of God. It would in the present state

of knowledge be next to impossible to give a clear definition
of such deep elementary principles, even if possessed by the
originators of the controversy, which they were not. Dif-
ficulties were increased by the doctrines of Arianism, a third
view of the same question. Arius was educated by Lucian
at Antioch. He appears to have rejected the theory of
" eternal generation," and would not accept a difference
between a generation from God and the notion of a creation.
In 321 A.D., Alexander, Bishop of Alexandria, excommuni-
cated Arius ; but Eusebius was an old friend, and inclined
to agree with him in doctrine. Constantine succeeding to
power in 324 A.D., found the Arian schism, regretted it,
and determined to unite all his subjects in one worship.
His words were, " πέρι μεν οὖν τῆς θείας προνοίας μία τις ἐν
ὑμῖν ἔστω πίστις." But his attempts to reconcile Alexander
and Arius failed, and a serious semi-political movement
broke out in Egypt between the Arian and Melitian party.
Constantine was puzzled how to act. He sided with
Alexander on the subject of the " unity of essence ;" but the
orientals objected that the doctrine gave occasion for sensu-
ous representations. He summoned the Council of Nice,
325 A.D., who as well as they could embodied the orthodox
belief in the Nicene Creed. Constantine deeming the Arian
tenets subversive of Christianity, used all the expedients
of Byzantine despotism in suppression ; he banished first
Arius, and then Eusebius.

It would be strange if no impress of these distracting
controversies by three powerful sections was left on the
holy places at Jerusalem ; surprising if there alone they
could unite on points which elsewhere defied the strength
of the Eastern Empire. The traces were left ; and it is not
easy to account for two distinct sites for the Holy Sepulchre,

one in the heart of the town, and the other on the Haram Area, without going back to the origin of these rifts in Christendom. It was as part of Constantine's scheme of unification of public worship, that he resolved to erect basilicas and memorials over the sacred sites. The Empress Helena went over to Palestine, and Constantine entered upon the operations at Jerusalem 326 A.D. Neander observes that, "Nothing certain is known with regard to the relations between Helena and her son as to this matter." But anyone who will examine such records as exist, will find that there was a complete divergence in their sentiments; and that the influences under which empress and son acted were wholly different.

Constantine's brother-in-law, Licinius, was a pagan, who however assumed the direction of Christian ritual in his part of the empire. He was the first who divided males and females in Christian congregations. Constantine, on the other hand, destroyed the pagan shrines. Among the ladies of Constantine's family there was a strong disposition to favour Arius. By their intercession Eusebius was restored to place in 327 A.D., and Arius was received two years later. Athanasius, Bishop of Alexandria at the time, was peremptorily commanded on pain of exile to ratify the measure. There is other confirmatory evidence of a genuine difference between the objects of Helena and Constantine. Doubts have been expressed regarding the correspondence, and there is no detailed account of what the empress did from first to last when at Jerusalem. Eusebius has a mere paragraph ("Vita," lib. iii., cap. xlii.), "For when she had decided to perform the office of pious affection due to Almighty God, both on behalf of her exalted son emperor to wit, and on behoof of her sons the Cæsars, dear to God, when she

had charged her grandchildren that they should pray with
supplications, although already of advanced age, yet with
juvenile mind, she, a woman of rare shrewdness, hastened
to travel through the land worthy of veneration; and she
visited the provinces of the East, and the cities, and people,
with an almost regal solicitude and preparation. But after
she revered, with appropriate worship, the steps of our
Saviour, according as the prophetic declaration had of old
predicted, 'Adoremus in loco ubi steterunt pedes ejus,'
forthwith also left the fruit of her piety to her descendants' (xliii.) ; "For she immediately dedicated two temples,
one at the cave in which the Lord was born, the other in
that mount from which He had ascended to heaven." There
is no mention of Helena having built any Church of the
Resurrection, but at the same time it is a matter of uncertainty on what mount she founded the Church of the
Ascension ; because a little further on in the same chapter
(xliii.), Eusebius says : "Besides that of the Ascension of
the Saviour of all to heaven, the mother of the emperor
raised a memorial, with lofty constructions, upon the
Mount of Olives, erecting the holy house of the church and
a temple over the summit of the whole mountain." The
ascension is known not to have taken place from the top of
the Mount of Olives, so that Helena cannot have been
accurately guided, if that event was commemorated by the
church and temple.

The Church of the Resurrection, however, was taken in
hand by Constantine himself, whose writings show that he
was as accurately informed on every point of Gospel
history as at this day we can be ourselves. He had
evidently experienced a difficulty in ascertaining the proper
site, but without having the communication in which the

invention of the cross, or perhaps more correctly the identification or indication of the scene of the passion of our Lord, was described, it would be impossible to say if the emperor relied upon the miracle when ordering the construction of the Martyrium. The rescript to Macarius, Bishop of Jerusalem at the time, is worth quoting:

"The Victor Constantine the Great and Worshipful to Macarius.

"So great is the grace of our Saviour, that no abundance of speech seems adequate to the narration of the present miracle. For the indication of that sacred passion, till now hid for such a long space of years under ground, through the base enemy of all, till its removal, to have appeared bright to the servants freed of it, really exceeds all admiration. For if all who have ever been considered wise on earth had been collected into one and this place, should wish to say anything worthy of this deed, they would not seem to me able to rival the least part of it. For the repute of the miracle as much surpasses as human divine things. Wherefore this is alone and especially my sole aim, that, just as the faith of the truth is daily exhibited with new wonders, so also our minds may be incited to observance of the most sacred law with all modesty and concordant alacrity. And because I think it manifest to all, and I wish that to be particularly credited, how it concerns me most of all, that the very holy spot, which by the providence of God I have relieved from the basest accompaniments of an idol, as if from a superincumbent weight, which had been holy from the beginning by the judgment of God, and has been manifested more holy from when the evidence of our Saviour's passion has been brought to light, we should adorn with the beauty of buildings. It becomes therefore your acumen that you should so arrange and care for the things necessary for the work, that not only should the Basilica be better than any everywhere, but that the rest be such that all the fairest things of the kind in other towns may be distanced. And concern-- ing the raising of the walls and adornment, to Dracilianus our friend, ruling the sections of the most important provinces, and to the Archon of the province, the care has been entrusted by us. For it has been ordered by our piety that workmen, and artificers, and all needed to obtain for the building, advised by your discretion, should through their foresight immediately be sent. However, regarding the marble columns you may think to be most valuable and useful, their synopsis drawn, hasten to write to us. That as many and of what sort through your letter we may know necessary, these can be passed over from all quarters. And the vault of the Basilica, whether fretted, or it strikes you should be by any other work, I wish to be told from you. For if you would prefer a fretted roof, it can be embellished with gold. The rest thy holiness will make known quickly to those foresaid judges, and how much labour, artisans, and treasury money; and that you may refer to me direct not only concerning the marbles and columns, but those things concerning the frets, if you think this work should be more splendid. May God pre- serve you, loved brother."

As Eusebius is so silent with regard to the circumstances of Helena's pilgrimage, it is useful to turn to any light, how- ever dim, that other authors can throw.

Sozomen lived about 400 A.D., or seventy odd years after

the erection of the Basilica. He says Helena tried to find the Holy Sepulchre. It was no easy matter. The Greeks, who at the first promulgation of Christianity tried to exterminate it, heaped up mounds of earth on the holy places, and had enclosed the place of the resurrection and Mount Calvary within a wall, and had ornamented the whole locality and paved it with stone. A temple and statue (ζωδιον, little animal) to Venus had also been erected on the same spot by these people, that the true cause of worship might be forgotten, and that the temple and statue would come to be regarded as exclusively appertaining to the Greeks. At length the secret was discovered, some say by a Jew. When the place was excavated the cave was discovered, and at no great distance crosses. Constantine erected a temple. Helena also erected two. Such is the substance of the narrative of Sozomen. A decided line is drawn between the structures erected by Helena and the emperor, agreeing with Eusebius, but supplying a few particulars in which his account is deficient. Abulfeda (" Hist. Ante Islam ") who lived 1200 A.D., is another authority, a Mohammedan viewing history from his own standpoint. He writes: "Urbs Hierosolymorum a Tito vastata instaurari cœpit. Aliquis Imperatorum Romanorum ei operam dedit. Eam Æliam appellari jussit quod significat domum Domini; eam restauravit et ædificia collapsa rearsit. Hæc tertia urbis instauratæ etas ad illud pertinet, quo Helena crucem quæsitura eo venit. Tunc enim sepulchro in quo christum jacuisse existimant ecclesiam quæ Kumamah vocatur superstruxit, templum solo coæquavit, et in loco quo steterat, urbis purgamenta et sordes conjici jussit. Ita locus es Sakhræ (Sacri Saxi in quo Jacobus dominus caput deposuisse dicitur) in sterquilinium conversus est. Sed," etc.

This is exactly what would be done under Constantine's orders, whose intention was to unify public worship, and obliterate any remnants from the Jewish ritual; although, had the emperor ever gone to Jerusalem, his taste would hardly have permitted the Temple site to be more than levelled. But the sentence last quoted reads against the theory that the Sepulchre Church of the Kumamah was near the Sakhra. Looking more closely at the quotation, it will be seen that church and temple are named in one breath, and may be fairly inferred adjacent. On the other hand, the expression "locus es Sakhra" does not necessarily mean the same as "es Sakhra" by itself, and it is quite within credibility that the premises of the Sakhra might be abandoned to defilement, whilst the Sakhra was built around, and preserved intact. The Temple that was demolished was no doubt that in which the image of Jupiter was put by Hadrian. The suspicion, however, gathers intensity that there was some kind of Christian establishment at that time upon the spot in the heart of the town, where there now stands the Church of the ;Holy Sepulchre. It may have been on no scale, but sufficient to affix importance to the locality. The section of Christians likely to occupy that situation would be the Nazarenes, who were objects of aversion to the Romans. Whether the cross was discovered here by Helena, and Constantine, learning the topography, disregarded the site as he appears to have done for his Basilica, and placed it on the Haram instead, are alike uncertainties. But at the time of Helena's visit the Arians were probably in occupation, having succeeded the Nazarenes, and the empress was under their control, and imbued with Arian sympathies. Arius himself may have been more unguarded than erring, but Blunt ("Dic. of Sects") observes: "If Arianism be true, Jesus Christ was not God. It necessarily leads to an open denial of the divinity of our Lord." There are no grounds for considering either Constantine's or the empress's letter spurious, and the contents amply account for his own interference with and taking in his own peremptory care the building of the Church of the Anastasis. They are as follows:

"To Victorious Triumphant Eternal August Son Constantine, Helen Eternal August.

"Wise reason does not reject truth, nor truly right faith ever sustain any damage. Therefore about to be deemed accepted by divine goodness, because you have been held approved who have receded from the vanity of idols; but about to lend yourself to error; because you have believed Jesus Christ to be true God, and this Son of God to be in the heavens, who a Jew, and on account of the theology of the Magians was condemned, and by sentence pronounced, of suffering the Cross, died. But to your piety so far was success accorded, because you first among emperors displaced the idols. For the true and eternal God wishing to show those whom you have denied, to be no true gods, has granted you health, by which having laid aside all fear of idols you can know, because neither asked can they confer safety or irritated take away any. Weakness has left thy vanishing reverence the error of this superstition; now virtue will attend thee returning to the omnipotent God, who never could be overcome. Which beginning to worship, you will obtain the Kingdom of David and of wise Solomon. For there are about to be with thee prophets to whom God spake, and whatever you will ask through them you will obtain."

Constantine's reply was:

" To Lady Eternal August Helen Mother, Son Constantine Eternal August.

" God who of all ages is ruler, who governs, directs and vivifies us all, through whose favour it is He bestows life and breath, and to all princes has deigned to entrust that they may give their rights to men. As therefore we have eminence among men, so far human hope centres on us. Therefore the eyes of all, the wills of all, what to us beseems, they incline to do; what we do not smile at, they shun. Wherefore, Lady Eternal August, not only irreprehensible but also commendable appears our pleasure; and really think nothing honest we do not wish. But what I have said is placed in our dispositions, but to know God is beyond our insight of mind. Wherefore cease the opposing of our rashness. But rather let the bishops of the Jews assemble into one, we ourselves hearing, and concerning these things dispute among themselves, that by their objections and arguments we may prevail to know the evidence and certainty of the true faith. For thus from Holy Scriptures we can, they and us, concerning the truth become assured, and encourage the whole world to right and firm faith. Adieu, Lady Mother Eternal August, florid with happy events."

The instructions given in Constantine's own words relate to a basilica at Jerusalem. Originally basilicas were halls of audience and justice, and the form, as accommodating a large number of people, was early adopted for Christian churches. A very good idea of what a basilica of those days was, can be formed from the longitudinal section of the Church of the Nativity at Bethlehem, which has been preserved, and is a work of Constantine's. If the Emperor's orders were carried out, the Basilica of the Resurrection must have been a still more imposing structure. Sozomen says: "περι παντα τον της αναστασεως χωρον, και του χρανιον διεκοσμησα. There was also a curious tradition, Chrysostom mentions, that Adam died in the place of a skull, and lies there. He cites Zech. xii. 10. But the authority for what Constantine actually built is Eusebius, who was a contemporary of the Emperor. His writings are important, because they are almost the only testimony we have, except such as can be drawn from the architectural character of any surviving remains. Much difficulty has been raised in the interpretation, as is to be expected when an author without technical knowledge undertakes to describe a group of buildings, without plans, in a distant country. The account given by Eusebus is however so critically important that an interlinear translation of Vita c, 33—40, is for the first time brought forward :

Και   δε   κατ   αυτο το σωτηριον   μαρτυριον   η
And indeed round that the Saviour's place of witness the
νηα κατεσκεναζετο Ιερουσαλημ αντιπροσωπος   τη   παλαι
new   Jerusalem was built   oppositely faced to the   old
βοωημενη,   η   μετα την   κυριοκτονον   μαιαιφονιαν
famed one which after the Lord's execution, slaughter pol-
ερημιας επ εσχατα περιτραπεισα, δικην   ετισε
luted desolation to last degree experienced the penalty paid
δυρσεβων   οικητορων ; ταυτη δε   ουν   αντιχρυς
of the impious inhabitants ; to this then therefore opposite,
βασιλευς την κατα του θανατου σωτηριον   νικην πλουσιαις
the king,   over   death the Saviour's victory with
και δαψιλεσιν ανυψου φιλοτιμιαις ταχα   που   ταυτην
rich and copious honours raised ; rapidly where that

ιυσαν την δια προφητικων θεσπισματων κεκηρυγμενην

was which by the divine Prophets had been proclaimed,

καινην και νεαν Ιερουσαλημ ης περι μακροι λογοι

fresh and new Jerusalem, which concerning long accounts,

μυρια δι ειθεου πνευματος θεσπιξοντες

a thousand things by the divine spirit prophesying,

ανυμνουσι. Και δη του παντος ωσπερ τινα κεφαλην

celebrated. And indeed of all as if in a sort the head,

πρωτον απαντων το ιερον αντρον εκοσμει, μνημα

first of all the Sacred Cave he adorns, the sepulchre

εκεινο θεσπεσιον παρ ω φως εξαστραπτων ποτε

so divine beside which light glancing, formerly

αγγελος την δια τον σωτηρες ενδεικνυμενην παλιγγενεσιαν

the angel that by the Saviour declared new generation

τοις πασιν ευηγγελιζετο.

to all announced.

Τουτο μεν ουν πρωτον ωσανει του παντος κεφαλην,

This then therefore first, as if of all the head,

εξαιρετοις κιοσι, κοσμω τε πλειστω

with selected columns, and with the greatest taste

κατεποικιλλεν η βασιλεως φιλοτιμια παντοιοις

was varied by the royal munificence with manifold orna-

καλλωπισμασι καταφαιδρυνουσα. Διεβαινε δ εξης

ments, making it resplendent. And one crossed next,

επι παμμεγεθη χωρον εις καθαρον αιθριον αναπεπταμενον

over a very great area, into a clean court open to the skies,

ον δε λιθος λαμπρος κατεστρωμενος επ εδαφους

and which a brilliant stone distributed about the paving-

εκοσμει μακροις περιδρομοις στοων εκ τρι-

stones decked, with long rounds of corridors from three

πλευρου περιεχομενον.

sides surrounded.

Τω γαρ καταντικρυ πλευρω   του αντρου   ο δε προς
For on the opposite   side   of the cave,   and that   to

ανισχοντα ηλιον   εωρα ο   βασιλευς συνηπτο   νεως
the rising sun which looks, the king connected the Temple,

εργον εξαισιον εις υψσος απειρον ηρμενον μηκους τε
a   work   to an untried height elevated, heights and

και πλατους επι πλειστον ευρυνομενον ου τα μεν εισω της
also widths to the farthest, extended: whose interior

οικοδομιας υλης   μαρμαρου ποικιλης   διεκαλυπτον
structural material, with variegated marble they concealed,

πλακωσεις   η δε εκτος των τοιχων οφις   ξεστω
plating it over, and the outer surface of the walls with

λιθω ταις   προς εκαστον αρμογαις συνημμενω λαμπρυνομενη
dressed stone, each to each, in bond adapted, resplendent,

υπερφυες τι χρημα καλλους της εκ μαρμαρου προσοψεως
admirable as a work, the fair aspect of a marble   surface

ουδεν αποδεον, παρειχεν.
wanting not, stood forth.

Ανω δε προς αυτοις οροφοις τα μεν εκτος δωματα
And above over those roofs, the domes that were outside

μολυβου περιεφραττεν υλη ομβρων ασφαλες ερυμα
also, with lead he protected ; a material, a safe preservative,

χειμεριων   τα δε της εισω στεγης γλυφαις
from wintry rains. And the interiors of the roof,   with

φατνωματων απηρτισμενα, και ωσπερ τι μεγα
carvings of the framings, was filled, and as   some great

πελαγος καθ' ολου του βασιλειου οικου   συνεχεσι ταις
sea down the whole of the royal house, with panels

προς αλληλας συμπλοκαις   ανευρευνομενα χρυσω
connected to each other again and again traced, and with

τε διαυγει δι ολου κεκαλυμμενα, φωτος   οια
burnished gold entirely   covered, of light as with the

μαρμαρογαις του παντα νεων εξαστραπτειν εποιει.
scintillations, the whole Temple he made to sparkle.

Αμφι δ' εκατερα τα πλευρα διττων στοων αναγειαν
And round each of the sides of the two corridors above,

τε και καταγειων διδυμοι παρασταδες, τω μηκει τον
and also below ground, double porticos, to the height of

νεω συνεξετεινοντο χρυσω και αυται τους οροφους
the Temple alike extended ; and with gold their roofs were

πεποικιλμεναι, ων αι μεν επι προσωπου του οικου
dotted, of them some on the front of the building

κιοσι παμμεγεθεσιν επηρειδοντο, αι δεισω των
by grand columns were supported, and those within in

εμπροσθεν υπο πεσσοις ανεγειροντο πολυν τον
front of these under soft gravel were built up much of the

εξωθεν περιθεβλημενοις κοσμον. Πυλαι δε
exterior ornament on those (columns) so girt. And gates

τρεις προς αυτον ανισχοντα ηλιον ευ διακειμεναι τα πληθη
three to the rising sun well disposed, the crowds

των εισω φερομενων υπεδεχοντο.
of those borne within, regulated.

Τουτων δάντικρυ το κεφαλαιον του παντος ημισφαιριου ην
And opposite these the head of all was the hemisphere,

επακρου του βασιλειου εκτεταμενον ο δε
from the extremity of the Basilica led off; and this

δυο και δεκα κιονες εστεφανουν τοις του σωτηρος
twelve columns adorned, to the apostles of the

αποστολοις ισαριθμοι κρατηρσι μεγιστοις εξ αργυρον
Saviour, equal in number, with great capitals made

πεποιημενοις τας κορυφας κοσμουμενοι ους δε βασιλευς
of silver, their heads ornamented, and which the king

αυτος αναθημα καλλιστον εποιειτο τω αυτου θεω.
himself made a beautiful offering to his God.

Ενθεν δε προιοντων   επι τας   προ του νεω
And of those proceeding within at the approaches lying
κειμενας εισοδους   αιθριον διελαμβανεν ησαν δε ενταυθοι παρ
before the Temple, a court intervened and were there near
εκατερα   και   αυλη   πρωτη   στοαι   τ' επι   ταυτη
each other also the first hall and corridors against this,
και επι πασιν   αι   αυλειοι πυλαι   μεθ' ας   επ' αυτης
and for   all   the entrance gates.   After which, in the
μεσης   πλατειας   αγορας   τα του   παντος
midst of the street of the market itself,   the places before
προ πυλαια φιλοκαλως   ησκημενα, τοις την   εκτος
the gates of the whole suitably planned,   to those entering
πορειαν ποιουμενοις καταπληκτικην παρειχον την των ενδον
from   without,   striking   exhibited, the prospect of
ορωμενων θεαν
those things visible within.

Such is the description of the buildings actually erected
in the Haram area by Constantine that Eusebius gives. The
interlinear translation requires and allows the meanings,
rather ruggedly disposed, to be corrected as far as accuracy
demands, and the Greek is more satisfactory than the more
rounded periods and necessarily foregone conclusion of the
Latin version. Briefly, Constantine raised a New Jerusalem
opposite the old guilty site, and began by decorating the
Sacred Cave with columns and ornaments. Before the
Cave was a very large open space, and beyond a paved
court open to the skies surrounded on three sides by long
colonnades. This open space was evidently to west of the
Cave, and not, as hitherto supposed, at the Golden Gate;
because on the opposite and eastern side the Emperor
connected the Cave with the Basilica, whose dimensions
were very large, the outside of ashlar, and the inte-

rior covered with variegated marble. The roof of a basilica would properly be of wood. There is no mention of timber in the description; so that the roof may have resembled the vaulting of the Golden Gate, and any domes were covered with lead. The expanse of the roof was in panels, like a vast sea, and highly gilt. This certainly suggests a modern roof with a flat pitch. There were side corridors in two stories, with porticoes the same height as the Basilica, their roof covered with ornament. Some of the ornamental roofs were supported by grand columns, the others by underground columns, while there were on the east three gates to regulate the crowd. Opposite these gates, leading off from the top end of the Basilica, was the hemisphere of twelve columns with silver capitals, the offering of Constantine himself. Between the entrance-gates and the Basilica a court intervened. There was first a hall, then against this corridors, and lastly came the gates. After which the pavements before the gates were skilfully set out in the middle of the thoroughfare of the market, and from the " propuleia " there was a striking view of the magnificence within. It is difficult to make more of these chapters of Eusebius than the sketch. The fortieth chapter of Eusebius distinctly states that the Emperor constructed this temple, the " white stone Marturion," as testimony of the Saviour's resurrection. Whatever may have been the intention, the Sacred Cave, although it may have been covered with a hemispherical dome, was only adorned with columns. Eusebius would scarcely have quoted the instructions to Macarius, or have followed this with such an imperfect description if he had ever seen the completed buildings. They agree in so far as this, that the orders are for a " Basilica et reliqua membra," and it was only the Basilica and its members that were erected. Eusebius when writing had no doubt access to information, and was guided in a literary way by this, so as to avoid decided inaccuracy; but the absence of local knowledge has caused the twenty-ninth chapter to differ from the letter to Macarius which immediately succeeds. For Eusebius says, that the Emperor forthwith directs a solid structure to be built of sacred model, with ample and regal magnificence, round the Cave of Salvation; οικον ευκτηριον θεοπρεπη αμφι το σωτηριον αντρον εγκελευεται πλουσια και βασιλικη δειμασθαι πολυτελεια; but when the only letter which exists is examined, it is not clear whether this sentence refers to some such edifice as the Dome of the Rock, or to the Basilica and its appurtenances, which was really put up.

The probability is that the Emperor originally proposed to erect a basilica, with a chapel all in one over the Cave; but that the Empress and those about her prevented the scheme being carried out in its original conception. The Basilica was placed near, but clear of the Cave, and a hemispherical colonnade, a pavement, and columnar decorations were erected actually round the sacred rock. In a sense, the whole group was around the Cave. But as there is the evidence of a strong party in Jerusalem at the time, it is unlikely that their opinions would not find some expression in the case of buildings whose sites comprised the points round which controversy revolved. After centuries are, at all events, indebted to this for the absence of particulars regarding the conclusion of the Emperor's correspondence with Helena. But there can be little question that the rock which contained the Cave was adorned by Constantine; and whether the building about it took the form of a circular colonnade of a kind seldom met with, or of a hemispherical dome over a set of twelve columns, depends very much upon the significance attached to the word hemisphere. It makes, indeed, very little difference; because if the Basilica of Constantine were in the Haram area, the nucleus of a building must have been placed by him, where the Dome of the Rock now stands.

As an example of the style in which the Basilica was built, we have the church at Bethlehem, an undoubted work of Constantine's time, to which to refer. The longitudinal section given by De Vogué contains a cave under the transept, suggestive, but at the same time inapplicable to the Basilica. The plan is a compromise between Byzantine and Latin features, and the low double rows of columns which support the clerestory upon a horizontal architrave specially draw attention, as the same trabeate construction is followed in the Dome of the Rock. So with the semicircular window heads.

The Basilica at Jerusalem bore a considerable resemblance, it may be supposed, to the church at Bethlehem. According to the scanty description of Eusebius it must have been more beautiful, coated with coloured marble inside, and its freshly dressed stone almost marble-white on the exterior. But the site was the sloping side of a hill, and the building lying right athwart the slope required to be adapted to its position. We have corridors on each side with a means of ascent to the main floor no doubt, and at their extremities double-storied porticoes, whose roofs were carried on grand columns; and inside were built under what seems to be concrete.

The Golden Gate must be the principal entrance of the new city, though outworks and a row of arches built into the Haram Wall are still unexplored and unexplained. If the Ordnance Survey Photographs (Captain Wilson's) are carefully examined it will be found that of the east elevation of the Golden Gate, all above the main cornice is stonework of the same age as the Damascus Gate, 1587 (?) A.D. The cornice is of fourth century work, wanting the upper projecting member, and may have been reset. The abutment of the curved cornice to north seems old work. The south abutment has been repaired and underpinned. There is again old work near the ground ; and two courses of the large rusticated stones of the Haram Wall appear to be carried along into the exposed basement of the Golden Gate. The Haram Wall above ground does not bond with the gate, but looks as if it were notched on. The inner or west face will be seen to be a complete architectural composition ; and unless a later copy, may be set down as fourth century work, remaining as it was built. The south side elevation has finished pilasters, and mere stone blocks for capitals ; but most of it appears to be equally ancient masonry. The roof, and perhaps the interior columns and the ceiling they support, are, as De Vogué considers, of Byzantine date, re-edifications of a dismantled portion.

There is something in the appearance of the east face to suggest that formerly there was an upper story ; and in the sunk condition of the structure that it was designed to conduct from a lower to a higher level.

The Church of St. George at Saloniki belongs to Constantine's age. Its construction is similar to the Pantheon at Rome, and the material is brick without mouldings. Nothing can be learnt from the appearance of the building, but a curious, though remote, correspondence subsists between the Mosaics on the interior of the Dome, beautifully drawn by Pullan and Texier, and the group of Constantine's edifices in the Haram Area as fancy combines with Eusebius to depict them. M. Texier observes: "Quoique la composition de l'architecture de ces tableaux soit variée, le sujet est toujours le même; il représente un petit temple au millieu d'une splendide colonnade." Since Constantine founded Saloniki in 326 A.D., and resided there for two years, erecting churches and other works, this set of mosaics may very readily have been pictorial outlines of the project he was contemplating and had actually begun for Jerusalem; that is, if they are as old as the Dome. The order of the Knights of the Holy Sepulchre is said to have been instituted by the Empress Helena in 302 A.D., but how far their prowess aided and maintained the work is unstated.

However, by 336 A.D., the Basilica at Jerusalem was completed; and the Emperor, desirous of consecrating the building in peace, summoned the Council of Tyre to decide the Arian controversy. Arius was received into the communion of the Church, and Athanasius was banished. Constantine died in 337 A.D., and twenty years after the Arian philosophy was triumphant in the West. In the East, Cyril, Bishop of Jerusalem, was deposed and banished. The testimony of Cyril, who was born 315 A.D., and actually preached in the Basilica which Constantine had erected, is very valuable. He writes, Cat. x.,

Αλαλαγμα   μονον·   επι   του   τυπικου

κατεπεσε   τα   τειχε της Ιεριχο, και   δια   το
there has fallen the walls of Jericho, and through the

ειπειν τον Ιησουν " ου   μη   αφεθη   ωδε λιθος επι
utterance of Jesus " there shall not be left here stone   on

λιθον,"   πεπτωκεν   ο   αντικρυς   ημων των Ιουδαιων
stone," there has fallen the opposite   us   Jewish

ναος ουχ οτ η αποφασις του πεσειν αιτια αλλ
temple, not that the utterance was the cause of falling, but

οτι η αμαρτια των παρανομων γεγονε του πεσειν
because the   evil   of transgressors was the cause of

αιτια.
falling.

Here Cyril talks of the Temple of the Jews as opposite where he was speaking. He proceeds, Cat. xi.:

Ο γολγοθας ο αγιος ουτος ο υπερανεστηκως
Golgotha the sacred   which   had stood high above,

μαρτυρει φαινομενος, το μνημα της αγιοτητος μαρ-
witnesses as it appears ; the sepulchre of   holiness   wit-

τυρει και ο λιθος ο μεχρι σημερον κειμενος. Cat. xiii.
nesses; and the stone up to this day lying.

Και γαρ αρνησομαι νυν ελεγχεν με ουτος ο
And were I sceptical, now would reprove me that very

γολγοθος ου πλησιον νυν παντες παρεσμεν. Ελεγχει
Golgotha not near now we all are at present. Reprove

με του σταυρου το ξυλον το κατα μικρον εντευθεν
me also the wood of the Cross, that in a short time hence

παση τη οικουμενη λοιπον διαδοθεν. Εσταυρωθη
to all the world, the relic was distributed. The Lord

ο κυριος ειληφας τας μαρτυριας ορας του γολγοθα τον τοπον.
was crucified.

Then Cyril goes on to observe :

Ζητουμεν δε γνωναι σαφως που τεθαπται.'
We seek then to know clearly where he had been buried.

The prophets reply (Isa. li. 1 ; Ecc. ii. 11 ; 1 Pet. ii. 6 ; Isa. xxviii. 16) :

Μισησον τους λεγοντας οτι κατα φαντασιαν εσταυρωθη.
Abhor those saying that in effigy he was crucified.

ει γαρ κατα φαντασιαν εσταυρωθε εκ σταυρου δε
for if in effigy he was crucified and from the cross

η σωτηρια και η σωτηρια φαντασια. Cat. xiii. *Ο γολγοθος*
salvation, then salvation a phantom. Golgotha

ουτος ο αγιος ο υπερανεστως και μεχρι σημερον
so holy that had been elevated and to this day

φαινομενος και δεικνυων μεχρι νυν οπως δια Χριστον αι
appearing, and showing till now how through Christ the

πετραι ερρανησαν το μνημα το πλησιον οπου ετεθη
rocks were poured on the near sepulchre where he was laid;

και ο επιτεθεις τη θυρα λιθος ο μεχρι σημερον
that was placed on the opening the very stone to this day

παρα τω μνημειω κειμενος. Cat. xiv. *Θελεις δε γνωναι*
beside the tomb lying. And you wish to know

τον τοπον (refers to Cant. vi. 10) *και ποθεν εγηγερται*
the place and whence the Saviour had

ο σωτηρ. Λεγει εν τοις Ασμασι των Ασματων ii. 10, *και*
been raised. It says in the Song of Songs and

εν τοις εξης (Ibid. 14). *Σκεπην της πετρας ειπε την*
in the following the cave of the rock; it says there

τοτε προ της θυρας του σωτηριου μνηματος ουσαν
then, before the entrance of the Saviour's tomb, was

σκεπην και εξαυτης της πετρας καθως συνηθες ενταυθα
a cave; and always at the rock, as usually, there

γινεσθαι προ των μνηματων λελαξευμενων νυν
exists before the sepulchres hewn in stone, for now

γαρ ου φαινεται. επειδη τοτε εξεκολαφθη το
it does not appear after once was chiselled away the

προσκεπασμα δια την παρουσαν ευκοσμιαν προ γαρ της
envelope through the present decoration, for prior to the

βασιλικης φιλοτιμιας του μνηματος σκεπη ην εμπροσθεν
royal enterprise the cave of the sepulchre was in front of
της πετρας. Αλλα που εστιν η πετρα η εχουσα την
the rock. But where is the rock that contained the
σκεπην αρα περι τα μεσα της πολεως κειται η περι
cave? Perhaps about the midst of the town it lies; or towards
τα τειχη και τα τελευτεια και ποτερον εν τοις αρχαιοις
the walls and the suburbs; and whether in the old
τειχεσιν εστιν η τοις υστερον προτειχισμασι λεγει
walls it is; or in the nether outworks; it says
τοινυν εν τοις Ασμασιν, "εν σκεπη της πετρας εχο-
accordingly in the Song of Songs.
μενα του προτειχισματος" (Cant. ii. 14).

"Judas eos ad illum deduxit. Tum vero Deus speciem
Christi in Judam ipsum transtulit quare hunc abreptum
verberarunt ... tum eum quem pro Christo habebant, in
crucem egerunt, etc., ut tradant Mattheus, etc." (Abulfœda).
This is probably the belief of the unreflecting section of
Mohammedans.

It may be remarked (when reading Cyril, that he is
acquainted with all the sites except the Holy Sepulchre;
and the passage above quoted shows that its identification
was as uncertain in 349 A.D. as at the present day. Cat. xv.
" Ερχεται δε," says Cyril, " ο Αντιχριστος τοτε οταν εν τω
ναω των Ιουδαιων λιθος επι λιθον μη μεινη." The Arabic account of the buildings at Jerusalem is very scanty, but it
confirms the hesitation of Cyril. Jalal Ud Din speaks of
the "Bab al Mukadas," consecrated house. "In the last
times shall be a general flight to it, and the Shekinah shall
be lifted upon high in this temple." That it is the most
beloved place. When the Greeks obtained possession of it

they built upon it a building as broad at the base as it was high in the sky, and gilded it with gold and silvered it with silver. It fell. They built a second; it fell. They built a third; it fell. An old man came and said: "All holiness has departed and been transferred to this other place ; I will therefore point out this as the place wherein to build the Church of the Resurrection." He cheated them. He commanded them to cut up the rock, and build on the place he ordered them. Thus they demolished the mosque, and carried away the columns and stones, and built therewith the Church of the Resurrection. Jalal also makes Omad say: "To return to the Sakhra, the Franks had built a church upon it . . . adorned it with images and candlesticks, and erected separately from the other buildings, a little chapel raised upon marble pillars. The Sakhra was hidden, being covered by the buildings on it. The Sultan commanded this coat of marble to be stripped off, and building taken to pieces. Thus the Sakhra was restored. Moreover, the Franks had cut off a piece from the Sakhra, and had carried it to Constantinople. A piece to the Russians."

Orientals write history in a different mode from the narrative style of the West, but there is every indication here of the Sakhra having at one time been built upon by the Franks. That subsequently they were misled into cutting off a portion of the Rock, demolished the mosque which perhaps should be read Basilica, and with the columns and stones, as Cyril hints, built in the heart of the town the Church of the Resurrection. Meantime a body was rising which has not been well understood, and of whose influence little notice has been taken, in considering the question of the Holy Places. The Nestorians were originally (see

Badger) founded by Addai and Mari of the seventy
apostles in Mesopotamia. Mari went to Cashgar and elsewhere, and died in 82 A.D. One of his successors was a
relative of the Virgin Mary, and a disciple, Aha d' Abhoi,
was consecrated at Jerusalem in the Church of the Resurrection about 205 A.D., which shows that Constantine's
buildings were not the first Marturion.

The Nestorian community, if any, therefore, were acquainted with the earliest Christian traditions. The first
germ of the persuasion under that name, however, was a
sermon preached before Nestorius in 428 A.D. He was a
Syrian, and Bishop of Constantinople. About the end of
the fourth century a singular female sect, the Collyridians,
existed in Arabia; so called from the κολλυρα, or cake,
they dedicated to the Virgin. The spread of Mariolatry in
the Church is attributed to this sect, and to the invocation
of the Virgin Nestorius was opposed. He was displaced
from his bishopric 431 A.D., by the Council of Ephesus.
The publication of the Theodosian Code in 438 A.D. caused
persecutions, not of Pagans, but of Christians against
Christians. Eutyches, the Archimandrite of Constantinople,
opposing Nestorian views, asserted that although two
natures existed in our Lord before His incarnation, afterwards there was only one: for which about 451 A.D. the
term " Monophysite" was used.

This controversy divided the Eastern and Western
Churches. Nestorius, consequently, was cut off from both.
Nestorianism becomes important, and ceases to belong to
mere Church history. It spread in Persia, and Chosroes
and his court considered Nestorianism their established
faith. How far this maintained its purity is a theological
question; but it may easily be imagined to have retained

its traditions. Osborn mentions, and it will be found elsewhere, that when a youth Mohammed was taken, about
581 A.D., in a caravan of merchants to Bosra in Syria, where
he was for some time under the teaching of a Nestorian
recluse. As this town, situated in the tribe of Manasseh,
beyond Jordan, is not a hundred miles away in a direct
line, there is every probability of Mohammed having at an
early period of his life personally visited Jerusalem. At all
events, for the Scriptural travesties that are to be found in
the Koran, the tuition of the Nestorians is one of the ac-
knowledged sources. It will in great measure account for
the divided allegiance paid to the Caabah at Mecca and the
Haram Area at Jerusalem ; a conflict between Arab incli-
nations and policy and correct tradition. As a matter of
fact, the Nestorians are indirectly the custodians of the
Sakhra and the Temple site at present, through the Mohammedans.

Before the growth and expansion of the Mohammedans,
the Byzantine emperors continued to protect the Christian
churches and communities at Jerusalem. Sects will be
found to be numerous, differing upon abstruse and essen-
tially unimportant theories, which are as difficult to explain
completely now as then. After the Council of Chalcedon,
451 A.D., the Monophysites separated from the Orthodox
Greek Church ; found sympathy with, and were patronised
by, the Mohammedans, to whose belief they partially ap-
proached ; and formed the sect of Copts, which is spread in
Egypt, and has a place of worship in the modern Church of
the Holy Sepulchre. They were, however, depressed in
Justinian's reign. This emperor succeeded to power 527 A.D.,
and his great aim was uniformity of public worship. The
decision of the Council of Chalcedon was favoured by the

emperor, but was in Egypt considered one in the Nestorians' interest. There was an insurrection of monks in Palestine 451—453 A.D., led by the Empress Eudoxia. After the sixth century the Monophysites became known as Jacobites, and were established in Egypt by the Mohammedans, because the Orthodox or Melchite party were representatives of the Greek Empire. The Armenians also inclined to the Monophysite principle, and have been long associated with the Turks.

Justinian tried and failed to unite Catholics and Monophysites. In the East he persuaded the Jews to acknowledge Christianity, and this was evidently the origin of his constructing fresh buildings in the Haram Area. Instead of obliging the Jews to worship in the Basilica of Constantine, he endeavoured to conciliate their national prejudices, doubtless by raising his Mary Church upon the edge of the made ground that had been disturbed so very inauspiciously by Julian. In a letter to Bishop Epiphanius, Justinian wrote: "When, therefore, on a former occasion we had found that certain aliens from the Holy Apostolic Church had followed the deception of the impious Nestorius and Eutyches, we promulgated our holy edict as your holiness also knows, whereby we checked the madness of the heretics." The impression left by Procopius's account of the Mary Church of Justinian, a translation of which is given in Williams's "Holy City," vol. ii. p. 369 et seq., is that the Mary Church was situated at the south-east corner of the Haram ; and for some reason, rather implied than expressed, great expense was incurred in raising the substructure. But there is a curious passage to the effect: "The place, however, being situated inland, at a distance from the sea, and fenced off with abrupt mountains on all sides, as I have described, rendered it difficult for the contrivers of the Temple to introduce columns from elsewhere. But as the emperor was distressed at the difficulty of the task, God showed a kind of stone in the nearest mountains well adapted for the purpose, whether it existed and was concealed previously, or was now created. In either case there is credibility," etc.

This certainly looks as if some other building, containing columns of the kind required for the interior of a church, had been despoiled on this occasion; and an emperor who was endeavouring to place all churches on the same level would not scruple to authorise such an expedient. It would indeed be a great assistance in deciding upon some of the facts of the topography of Jerusalem, if the materials of the different buildings could be traced. It is a small place, entirely isolated in those days from large supplies of skilled labour, and having no roads. Therefore there was every inducement to make any decorative materials to hand available for a new design. The Palestine explorations have as yet shown no spare columns or useful squared stones, or even fragments to form part of the accumulations that have filled the valleys in Jerusalem. Yet the stones of the old temples and palaces of the Jews, the decorations of the Sakhra, and the Basilica of Constantine, and those of Justinian's edifices, must all. exist in some shape, and have been passed from one set of constructions to another. Justinian's church (see Tobler) was visited in 808 A.D., and was in all its glory when Bernhard was at Jerusalem in 870 A.D. So that it must have been for long standing by the side of the Mosque el Aksa. Procopius mentions that Faustinus, a man of Jewish extraction, but who for security had taken the name of a Christian, came to be governor of Palestine under Justinian. He was accused by the Jewish priests for his duplicity, and for having committed cruelties against the Christians during his government, and condemned at Constantinople. But by degrees a good sum of money mollified the emperor, and his credit became so great that he had unchecked management of the Imperial domains in Phœnicia and Palestine. Such an ally must have been a great assistance in dealing with the Jewish population, and promoted an outward, but probably fallacious, uniformity.

Justinian died 565 A.D. The various forms of worship centred at Jerusalem continued. We are now approaching an important period, when the Mohammedan creed was about being promulgated, and the Pope was on the eve of acquiring temporal power. The native Syrians were to a great extent Nestorians at this time; and Nestorians were opposed to the Greeks. Heraclius, the Greek emperor, from political motives, persecuted the Jews. So that when Chosroes II., the head of the Nestorian community—for Nestorianism was in 500 A.D. the established religion of Persia—led a revolt of the Persians, and invaded Palestine, he was accompanied by a large number of Jews, and the attack was made by two combined parties having each their own ground of dissatisfaction with the Greek Empire. Finlay relates that Chosroes burnt the Church of the Sepulchre; Williams states that, accompanied by Jews, he demolished the Church of Gethsemane, the Basilica of Constantine, the Churches of Calvary and the Holy Sepulchre. Why the Nestorian Chosroes should permit the destruction of the Basilica erected by Constantine, for a purpose they of all the sects might be supposed to revere, is involved in obscurity. But if the Jews set the Basilica on fire, it is intelligible that the Nestorians demolished the

Monophysite edifices of Gethsemane, Calvary, and the Sepulchre, that owed their commencement to the Arian party and the Empress Helena.

This period is marked in England, in a manner linking on to our times by the founding 611 A.D. of Westminster Abbey. But whether the churches at Jerusalem were more than damaged on this occasion is doubtful; as Williams states that in 937 A.D. the Mohammedans attacked the Church of Constantine, and laid waste the Churches of Calvary and the Resurrection. In 629 A.D. Heraclius retook Jerusalem, and as he endeavoured to reunite religions it may be conjectured that repairs of churches, as far as practicable, generally took place, and it is now we hear of Modestus having rebuilt the Church of the Sepulchre. Mohammedanism had now declared its principles. At Mohammed's uprising, Nejd, the central province of Arabia, under Moseylema, opposed his pretensions. It was bounded on the north by the Byzantine Empire, on the south by Yeman, and on the remaining land side by Persia. The inhabitants had socialistic proclivities, and held the same opinions which were more actively put in force subsequently by the Ismalians and Carmathians. Moseylema was slain at Hanefah by Kalid, after a fierce encounter, and his sect was dispersed, but it left Mohammed undisputed master, and an extreme dislike of Islam in the breast of every Arab attached to Nejd and its esoteric mysteries.

The downfall of this seemingly obscure sect placed Mohammed in an opposition that still subsists. Heraclius was at Jerusalem celebrating the restoration of the Holy Cross, which Chosroes had removed, and driving the Jews out of the city, when the first hostilities between the Mussulman and Roman troops occurred. At this time Sophronius was

head of a Greek or Melchite congregation, in the midst of
a hostile Monophysite or Jacobite population. This alone
would tend to mark the Haram Area to be his residence,
and any building of Constantine's remaining, as his church.
The last to be heard of Heraclius is when he mounted a
hill (Abulfeda Ams), and turning towards Syria, said:
" Vale Syria et ultima vale, neque enim mihi licebit dein-
ceps te invisere, neque Romano cuiquam te intrare, nisi
parenti, donea tandem nascatur inauspicatus ille infans,
quem nunquam nasci majis expediebat et optandum erat.
Tam ille Romanam rem insigniter affliget; tantos ille tan-
que acerbos tumultus excitabit."

In 637 A.D. Omar advanced to Jerusalem. The Christian
Syrians fled to Lebanon, and their descendants are known
as Mardaiites, supposed to be the present sect of Maronites.
Ockley's account is that Omar went first to the Temple of
the Resurrection. Then Sophronius took him out from
thence into Constantine's church, but he would not pray
there. Then he went alone to the steps near the east gate
of Constantine's church, and prayed there. Omar, leaving
the churches to the Christians, built a new temple where
Solomon's formerly stood. Notwithstanding his precau-
tions, the Saracens seized the church at Bethlehem, and so
they did St. Constantine's church at Jerusalem; for they
took half the porch where those steps were which Omar
had prayed upon, and built a mosque there, in which they
included those steps; and had Omar said his prayers in
the body of the church they would have taken that too.
Abulfeda relates in Latin: "The Locus es Sakhra was
turned to a dunghill, but things were altered no sooner
had Omar ibn el Chattal acquired Jerusalem. For he
having been shown by somebody the place of the Temple,

ordered it first to be cleared of filth, thereupon he built
there a Mohammedan structure, which remained uninjured
to the time of Abd el Malek, who having demolished that
building, placed on the old foundation another, which is
called the Musjid el Aksa; for whom the Sakhra was enclosed (cui inclusa est es Sakhra). In these no change
has taken place to our day, 1200 A.D. These are the matters
to which El Aris refers, who is responsible for the state-
ments. In which I think the destruction of the Temple of
Jerusalem he speaks about relates to that edifice which had
been built over the Sakhra": "nam Aedis illius El Mesjid
El Aksa jam in traditione sacra de adscensu prophetæ nos-
tri in cœlos mentio fit."

It is necessarily difficult to assign from Abulfeda's account their proper situation to churches and mosques, when
the narrator had himself such a confused idea of the real
circumstances. The whole would be consistent if the words
" Locus es Sakhra " and " cui inclusa est " did not occur;
but if the terms mean that the locality or near neighbourhood of the Sakhra, and that part of it where the Temple
of the Jews had been built especially, and not the Sakhra
itself, was cleared by Omar; and that he reared a mosque,
which was destroyed by Abd el Malek, so as to erect another on the same foundations, the same Abd el Malek who
appropriated and roofed the Sakhra, ambiguity disappears.
Finlay wonders how Sophronius, who was a Melchite Patriarch, consented to become the minister of the Mohammedans. But there is a probable cause, and Omar was
evidently averse to interfering with the Christian system,
working in harmony both in Constantine's and Justinian's
buildings on the Haram Area. There is all along an impression that the Monophysite party had separate religious

establishments, on a different site; to which, in order to confer the necessary sacredness, the fragments of the Sepulchre and perhaps a portion of the rock of Golgotha had been conveyed. If this was not the case, the Basilica of Constantine must have been repaired after its injury on the occupation of Chosroes. Because Ockley asserts that the " Sultan ordered bolts to be fixed to the Church of the Resurrection, and pilgrimage forbidden. Some advised it to be destroyed. The majority said that was no use ; they adore the site of the Cross and the tomb, not the buildings."

Omar cannot be imagined to have been unaware of the peculiar significance of the Sakhra, which has descended in the form of floating tradition rather than in precise terms; he however rested content with a small mosque. It is difficult to form an idea from Mohammedan accounts of their reasons for venerating the Sakhra. The subject is ably treated by Edouard Sayons, who brings out the fact that the Mohammedans are divided on the cardinal doctrine which is the base of the faith of Christendom. " Les commentateurs," observes Sayons, " ont peiné pour arranger tout cela, mais ils sont arrives à deux solutions différentes; d'àpres les uns Jésus serait monté au ciel sans passer par le sépulchre, d'apres les autres il aurait fait un court séjour dans le tombeau, il y serait resté trois heures ou sept heures."

It is generally asserted that the Sakhra derives its importance from having been Mohammed's point of departure on his night journey to the heavens; but this explains little, as the tale is obviously vague and allegorical. Jalal ud Din calls the Sakhra "the halting-place of the night journey, the resting-place of the Lord of Apostles, the resting-place of apostles and prophets, the mansion of Abraham." The preacher Kudi Moh is quoted as having said : "This is the spot from which your prophet ascended to heaven, this is the Kiblah, this is the reposing-spot of the prophets, this is the burial-place of the apostles, here descended the revelation, upon this land will take place the resurrection ;" and uttered a prayer, "that as entrance had been given into the consecrated Temple, they might be given entrance into the remaining parts of the land, and be made masters of the fortunes of the infidels and of their chieftains." The Arab song runs :

> "Great is my love ;
> If my love were in the Sakhra,
> That great and wonderful
> Rock the Sakhra,
> It would be broken
> Into a thousand pieces."

The history of Islam at this time becomes very much involved, and is the more difficult to unravel, because Mohammedanism is usually regarded as a unified system. Abu Sophian had long been the inveterate enemy of Mohammed, and 661 A.D. his son Moawiyah ascended the throne established by the prophet, and transferred the seat of government to Damascus. The Caliph Ali perished in the conflict, and the Suni Mohammedans gained the upper hand. The overthrown sect appears to have included opinions which afterwards were developed into those of the Ismaliens. According to them an Imam must belong to the house of Ali ; the world is never without an Imam, but he is not always visible. When visible, the doctrine is concealed; when hid, missionary labours begin; Prophets reveal, Imams interpret. In 684 A.D. Abd el Malek, the Ommyad, succeeded to the Caliphate at Damascus. There were four aspiring parties in the Mohammedan world at the time, and the principal opponent was Abdalla ibn Tobeir, grandson of the Caliph Abu Bekr, and nephew of the favourite wife of the prophet. It is related by Ockley, that Abdalla holding out at Mecca, " Abd el Malek enlarged the Temple of Jerusalem so as to take the stone into the body of the church, and the people began to make their pilgrimage thither."

The plan of the Aksa suggests that it was an enlargement of the mosque at the south wall built by the Caliph Omar, and the longitudinal section given by De Vogué, that the constructor had been guided by recollection, at all events, of the Basilica of Constantine, the flat architraves of which were conveniently replaced by Gothic arches. But the Aksa might have been quite independent of alterations of Constantine's remaining buildings, then probably verging on a ruinous state; and to promote the objects of pilgrimage he may have, as the Cufic inscription in the Cubbet es Sakhra asserts, "built this dome in the year 72 " (691 A.D.), independently of the Mosque el Aksa. Before the time of Abd el Malek, the Ommeyide Caliphs employed a divan of Syrians to conduct public business, and the records were in Greek. Arab officials and the Arabic language were his introduction, which will account for the subversion of the system that had been allowed to be perpetuated under Omar and Moawiyah. To some extent a divided Islam was due to the separate derivation of the Arabian tribes. The Koreish, to which Mohammed belonged, were descended from Ishmael; but there was another race of Arabs deriving their descent from an older Semitic stock, the Joktan Arabs; and the latter were the ruling masses at Damascus. The Caliph Abd el Malek died 705. At this period the Mohammedans had, it is probable, appropriated the Dome of the Rock, but allowed the peaceable occupation of Justinian's buildings, and any of the Basilica that was tenantable, as well as such establishments as the Christian community possessed in the heart of the town.

It was under the Fatimite dynasty that this toleration was first disturbed. About 900 A.D. Abou Abdallah was the Ismalien missionary for the son of the seventh concealed Imam. By giving himself out as the precursor of the " Mehdi," or Mohammedan " expected one," he was instrumental in the conquest of Kairwan in Northern Africa; and the result was the defeat of the native race and subjection of Egypt successively to Al Moer and the Caliph Hakim. The possession of Egypt by the Fatimites was, however, to be disputed by a sect founded on the principles of Abdallah, the fourth concealed Imam of the Ismalian line. He cast aside all theological beliefs, and rested human life and society on a basis of materialism. Carmath was a missionary of the sect who have been styled Carmathians. They increased in power and audacity, and in 937 A.D. sacked Mecca, tore up the pavement of the Kaaba, and split the black stone. No Moslems before or since have so greatly outraged Islam; but about the close of the century they were driven to the shores of the Persian Gulf, " a district," as Gifford Palgrave says, " a heap of exoteric doctrines." " The Wahabee reigns supreme there, but the Carmathian reaction burns secretly on, and waits but an occasion to break out afresh into a blaze sufficient to consume perhaps for the last time the superstructure of Wahabeeism and Islam."

An attack of the Mohammedans is recorded at this time by Williams upon the Christian edifices in Jerusalem. The Carmathian seizure of Mecca made it more than ever necessary to render the centre of pilgrimage in Palestine attractive, and whether they injured the Church of Constantine, and laid waste the Churches of Calvary and the Resurrection or not, we may be tolerably certain that at this time the Christians were rigidly excluded from the Haram Area. The belief of the Fatimites was that it was not enough for the Mohammedan world to have an infallible book, but there must be an infallible interpreter; and this is a knowledge which can only be inherited by right of blood from the Prophet. "It was," remarks Osborn, "under the Khalifate of Hakem, the grandson of the conqueror of Egypt, that these doctrines attained their fullest expansion." Utterly convinced of his own impeccability, he indulged every fantastic whim. But Hakem organised the constitution of his sect, and held "Conferences of Wisdom" in the palace. It is singular that at the present moment the Druses of the Lebanon regard El Hakem as their incarnate prophet, and believe that he will appear again. The Druses oppose the Turks, but incline to the Shiite faction, differing from them by their allegorical interpretation of the Koran.

In 996 A.D. El Hakem, nephew of the Patriarch of Jerusalem on the mother's side (De Vogué), incited by the Jews, demolished the Christian buildings. Lord Carnarvon states that the reason of Hakem's attack was rage at the information brought to him of the Greek fire, and the chain on which, oil being poured, the flame rose in the Church of the Sepulchre. It would appear that Justinian's buildings had already been destroyed, no doubt to effectually dislodge Latin Christianity from the Haram Area; so that the edifices most obnoxious to El Hakem's interference were the Dome of the Rock and the group of churches in the heart of the town; unless, indeed, Constantine's Basilica was used for the Greek fire. But it is to be gathered from such fragmentary accounts as exist, that the churches in the town were those attacked. For the Jews must have observed that these were the only link attaching the Christians to the Holy City. Hakem about one year afterwards restored the materials, and tolerated the Christians.

The present Church of the Holy Sepulchre was reconstructed by Greek architects in 1048 A.D., to become eventually the Crusaders' Church. The severe exactions of the Mohammedans furnished a pretext for the First Crusade; Jerusalem was taken 1099 A.D.; and, says De Vogué, "After their final victory the Crusaders only found in the Holy City the Church of the Resurrection, the Latin Convent of St. Mary, and the Basilica of Bethlehem. But there were other buildings, for Addison states that when the Crusaders came in 1099 A.D. they tore down the crescent from the Dome of the Rock, and put up a massive Cross. Isac. De Vibi, Hist. Hier., mentions that the Saracens held this building in such veneration that none dared to defile it. William of Tyre also speaks of the Octagonal Temple. The rock was left uncovered for fifteen years, after which the Crusaders cased it with an altar, on which priests officiated. The order of the Templars were now on Mount Moriah, their church was the Cubbet es Sakhra, their banner was a Lamb, and they lodged in the Mosque el Aksa. The entry of Baldwin is described by Guizot. In the morning the clergy and the king went to the Temple of the Lord, where Solomon's wisdom had been promised. The Greeks and the Syrians remained in the monument of the Holy Sepulchre. "Our people having prayed, returned to the Church of the Sepulchre."

In the eleventh century another body of Christians had settled near this church, who in 1130 A.D. became the military order of the Knights Hospitallers. Much about this time these very knights were presumably the Crusaders who are said to have reunited in one monument all the old isolated Christian sanctuaries, with evident reference to the churches in the heart of the town alone. In 1187 A.D. Saladin chased the Francs from Jerusalem, and only spared the Churches of the Holy Sepulchre, Josaphat, and Bethlehem; converting the rest to mosques or destroying them.

The disposition of the Saracen mind at the time is well illustrated by a sermon preached by a Mohammedan priest on Moriah upon Saladin's capture of Jerusalem, 1187 A.D. Extract: "Praise be to God, Who hath glorified Islamism; by His power hath debased Polytheism; I praise Him for having purified the polluted house from the impieties of Polytheism. Oh, men, publish the blessing, recapture and deliverance of this city which we had lost, and has made it the centre of Islamism after having been during one hundred years in the hands of the infidels. This house was built and its foundations laid for the glory of God and in the fear of heaven. For this house is the dwelling of Abraham; the ladder of your Prophet; the Kiblah to which you prayed at the commandment of Islamism; the abode of prophets; the aim of saints; the place of revelation; the habitation of order and defence. It was in this mosque that Mohammed prayed. It was this city to which God sent His servant, his messenger, the word which He sent to Mary," etc.

The Crusades were the cause of the temporary eclipse of the Greek Empire by the Latin. Baldwin took Constantinople in 1204 A.D., and in 1229 A.D. Jerusalem was again occupied by the Crusaders, representing a Latin dynasty; and who appear to have made slight restorations of the buildings. The treaty made on this occasion by Frederick, grandson of Barbarossa, with the Syrian Sultan, lasted ten years. It was broken, and the Mohammedans, who seem all the time to have reserved the Dome of the Rock for themselves, regained ascendancy. A fresh treaty, made in 1240 A.D., only lasted two years. Again in 1243 A.D. the Templars retook Jerusalem, but the Charesmians, flying before the Mogul power of Central Asia, overran Syria and Palestine; and in 1244 A.D. made themselves masters of the Holy City. The Charesmians probably reduced the Haram to a state of disrepair, and damaged the Christian churches.

St. Louis, King of France, was at Jerusalem in 1252 A.D., and resided there for four years, effecting nothing of moment. The Christians were always divided among themselves, the knights were moved to Rhodes and Malta, and Jerusalem became an easy prey, with the rest of Syria, to the Turks. The invention of firearms altered the conditions of defence, the Greek fires were completely paled, and in 1453 A.D., a memorable year, Mohammed II. took Constantinople, and the Turkish power assumed guard over the Haram Area.

A. T. FRASER.

December 6, 1880.

THE END.

BILLING AND SONS, PRINTERS AND ELECTROTYPERS, GUILDFORD.

www.ingramcontent.com/pod-product-compliance
Lightning Source LLC
Chambersburg PA
CBHW021526090426
42739CB00007B/806